Strawberry Shortcake Collectibles
An Unauthorized Handbook and Price Guide

Jan Lindenberger

D1797030

Schiffer Publishing Ltd

4880 Lower Valley Road, Atglen, PA 19310

Designed by Laurie A. Smucker

Typeset in Times New Roman 9/10 and 11/12.5
ISBN: 0-7643-0517-4
Printed in China
1 2 3 4

Published by Schiffer Publishing Ltd.
4880 Lower Valley Road
Atglen, PA 19310
Phone: (610) 593-1777; Fax: (610) 593-2002
E-mail: schifferbk@aol.com
Please write for a free catalog.
This book may be purchased from the publisher.
Please include $3.95 for shipping.
Try your bookstore first.

We are interested in hearing from authors
with book ideas on related subjects.

Contents

Credits

A very special thank you to Jolie Dickinson for her patience and kindness, and for welcoming me into her home so I could photograph her vast collection. The information she contributed was really appreciated and very valuable to this book. She would like to be in touch with other avid Strawberry Shortcake collectors. Jolie Dickinson can be contacted at 10707 Lake Creek Parkway Apt. #115, Austin, Texas 78750

Many thanks also to Jennifer Bowles, for supplying me with information from her overwhelming collection of Strawberry Shortcake. Jennifer publishes a newsletter about Strawberry Shortcake and her friends and welcomes any correspondence from other collectors. Jennifer buys and sells as well. For information on her newsletter you may contact her at "Strawberryland Gazette" 138 E. Main, Greenville, Kentucky 42345 or by phone at (502) 338-4318

I am also grateful to Chris Hillhouse from Erie, Pennsylvania who allowed me to photograph her collection, and to Melodie Pittman and Andy Comer from the T.C. Antique and Toy Mall in Pataskala, Ohio. They allowed me to photograph the many Strawberry Shortcake items from the mall.

"Baby Zoe" Angeni Dickinson-Ortiz, Daughter of Jolie Dickinson, cuddled up on a Strawberry Shortcake blanket.

Introduction

"Welcome to the World of Strawberry Shortcake," a world that American Greetings, Corp. created for little girls everywhere in 1979 and 1980. The Strawberry Shortcake character was originally created as a greeting card design. Little did they know it would be such a huge success.

Her world started small, but over the next five years it grew considerably. After all, who could resist her "berry" sweet personality? From 1979 to 1985 Strawberryland took hold of all of us with a new fascination on life from Strawberry Shortcake's view, and we all wanted to play in their sweet smelling world!

Now Strawberry Shortcake's world consists of the young and old alike, who want to enjoy Strawberryland and all its inhabitants!

"In a wonderful, imaginary place called Strawberryland, where strawberries grow all year long, lives a very lovable little girl called 'Strawberry Shortcake.'" This description came from a booklet of the Strawberry Shortcake toy line, produced by Kenner in 1980, and love her we did! Over the years, Strawberry took trips to "Big Apple City, China Cup, Mexicoco, etc. (as seen in the Strawberry Shortcake videos). It was on Strawberry Shortcake's trips around the world where she met other "berry" good friends who came to live in Strawberryland. Not everyone that Strawberry Shortcake met came back to Strawberryland or was made into a doll.

Kenner manufactured the dolls and accessories from the "berry" beginning. The original nine dolls were produced in 1980: Strawberry Shortcake, Orange Blossom, Blueberry Muffin, Huckleberry Pie, Lemon Meringue, Raspberry Tart, Apricot with Hopsalot Bunny, Apple Dumplin with Tea Time Turtle, Purple Pieman with Berry Bird. They were all scented to smell like their names. In all, there were nearly 50 dolls made (rag dolls, baby dolls, six inch dolls, etc.)

Starting in 1981, Kenner made all dolls with pets. In all, the "regular" line consisted of 19 dolls (five of which were International Friends, mentioned later in the introduction):

Angel Cake with Souffle
Apple Dumplin with Tea Time Turtle
Apricot with Hopsalot
Blueberry Muffin with Cheesecake
Butter Cookie with Jelly Bear
Cherry Cuddler with Gooseberry
Huckleberry Pie with Pupcake
Lemon Meringue with Frappé
Lime Chiffon with Parfait Parrot
Orange Blossom with Marmalade
Purple Pieman with Berry Bird
Raspberry Tart with Rhubarb

Sour Grapes with Dregs
Strawberry Shortcake with Custard

The other toys in the line were:

Berry Bake Shop
Strawberry Shortcake Berry cycle
Flitter-Bit the Butterfly
The Big Berry Trolley
Strawberry Shortcake Gazebo
Berry Merry Worm

From 1980 to 1984, 64 miniatures were featured as well and were all scented to smell like their names. 54 are fairly common:

Almond Tea with Panda in cart
Almond Tea with lantern
Angel Cake on the phone
Angel Cake in bubble bath
Angel Cake with Souffle
Apple Dumplin riding Tea Time Turtle
Apple Dumplin on sled
Apple Dumplin pulling a wagon
Apricot with Hopsalot in wheelbarrow
Apricot dancing with Hopsalot
Apricot with wheelbarrow
Blueberry and Cheesecake
Blueberry Muffin with hoe
Blueberry Muffin with basket of berries
Butter Cookie and Jelly Bear
Butter Cookie with Jelly Bear in a buggy
Cafe Ole with guitar
Cafe Ole with Burrito
Cherry Cuddler with Gooseberry
Cherry Cuddler on rocking horse
Crepe Suzette with stack of crepes
Crepe Suzette with Eclair Poodle
Huckleberry Pie and Pupcake
Huckleberry Pie fishing
Lemon Meringue with Frappé Frog.
Lemon Meringue picking a flower
Lemon Meringue looking into mirror
Lemon Meringue holding flowers
Lime Chiffon and balloons
Lime Chiffon the ballerina
Lime Chiffon with Parfait Parrot

Mint Tulip with Marsh Mallard
Mint Tulip with shovel
Orange Blossom and Marmalade
Orange Blossom with paint brush
Purple Pieman with Berry Bird
Purple Pieman holding pipe
Raspberry tart with Tasty Sundae
Raspberry Tart on roller skates
Raspberry Tart with Rhubarb monkey on see saw
Raspberry Tart with bowl of berries
Sour Grapes and Dreggs
Sour Grapes holding grapes
Strawberry Shortcake holding 3 berries
Strawberry Shortcake with bushel basket
Strawberry Shortcake on a skateboard
Strawberry Shortcake with her broom
Strawberry Shortcake with wrapping paper
Strawberry Shortcake with Custard
Strawberry Shortcake picking berries
Strawberry Shortcake in nightgown
Strawberry Shortcake with watering can
Strawberry Shortcake with birthday cake
Strawberry Shortcake playing with Custard

Another 10 minis made in 1984 are harder to find:

Almond Tea sitting on a cushion
Cafe Ole and Burrito Donkey dancing around hat
Cherry Cuddler with Gooseberry holding balloons.
Lem and Ada with Sugar Woofer Dog
Mint Tulip with Marsh Mallard duck in shoe bed
Peach Blush holding fan
Peach Blush in night gown holding mirror
Plum Pudding in nightgown with Elderberry Owl
Plum Pudding with Elderberry owl holding pencil
Strawberry Shortcake on tricycle

Deluxe minis are:

Almond Tea at table
Angel Cake at desk
Apricot with Hopsalot at vanity
Blueberry Muffin playing piano
Butter Cookie with Jelly Bear on swing
Cafe Ole at pottery wheel
Cherry Cuddler in airplane

Crepe Suzette with stove
Lemon Meringue at sewing machine
Lime Chiffon dancing at mirror
Mint Tulip with flower cart
Orange Blossom painting at easel
Peach Blush under trellis.
Plum Pudding with blackboard
Raspberry Tart in a car
Strawberry Shortcake in rocking chair
Strawberry Shortcake on sailboat
Strawberry Shortcake on a rocking horse

Kenner then produced toys for the miniatures as well. They were:

Berry Busy Bug
Strawberry Shortcake's House
Blueberry's Garden Shoppe
Raspberry's Soda Shoppe
Berry Patch Carrying Case
Storybook Carrying Case

In 1983, Kenner added new characters along with their existing Strawberry Shortcake dolls. They were presented as "Strawberry's Friends from Around the World." The international dolls and pets were named:

Almond Tea with Marza Panda
Cafe Ole with Burrito
Crepe Suzette with Eclair
Lem and Ada with Sugar Woofer (twins)
Mint Tulip with Marsh Mallard

Kenner also made a wonderful house for Strawberry Shortcake and her friends. It is called the "Strawberry Shortcake Berry Happy Home." It is three stories high and decorated daintily with strawberry shaped windows, a deck, a sunroof shaped like a strawberry, and a yellow porch swing. Many accessories were made for this doll house which included add-on furniture and deluxe furniture sets.

Additional dolls called "Party Pleasers" were introduced to Strawberryland in 1984. 10 dolls were introduced in this line and had clothes to match. They included:

Almond Tea
Angel Cake
Apple Dumplin
Cafe Ole

8

Cherry Cuddler
Mint Tulip
Orange Blossom
Strawberry Shortcake

and two new dolls:

Peach Blush with Melonie Belle lamb
Plum Pudding with Elderberry owl

Also in 1984 five Sweet Sleeper dolls were introduced. They had eyes that opened and closed, and came with sleeping bags. The boxes were made to resemble little bedrooms and included pets.

All good things must come to an end. In 1985, the final line of Kenner dolls, called the "Berrykins" by collectors, were produced. These six dolls had long hair that flowed to their feet and fancier outfits. They came with a "critter" instead of a pet. A new creation included in the Berrykin series was "Banana Twirl and her Banana Berrykin." They also had a Princess to guide them and keep them out of harms way. The Berry Princess was similar to a fashion doll with long blond hair and a beautiful pink and white dress. What's a "Berrykin" look like you ask? Well, it is kind of a cross between an elf and a berry.

In 1991, a company called Toy Headquarters (THQ) obtained the rights to produce the new Strawberry Shortcake dolls. They released nine dolls which had a 1990s look, with long styled hair, larger eyes, and updated outfits. Four of these were Strawberry Shortcake in various outfits; the others were Lime Chiffon, Orange Blossom, Lemon Meringue, Blueberry Muffin, and Raspberry Tart. These dolls were not as popular as the originals, but are still loved by most collectors.

Other rag dolls and accessories were also made. The 5-1/2" dolls came on large cards, not boxes, with extra outfits. No pets were included in this line. The line was discontinued due to lack of interest. The bakery that made the "Little Debbie" line ran a promotion for people who bought Little Debbie Strawberry Shortcakes offering them a "classic Strawberry Shortcake" doll. This doll was essentially a THQ doll packaged in a special box with wrist tag.

In 1997 American Greetings released stickers and greeting cards with Strawberry Shortcake's picture, but no dolls. So many companies held licenses to produce Strawberry Shortcake items, such as bedding, stickers, glassware, etc. that it is clear why Kenner once referred to it as "Pink Gold."

So why did they stop making Strawberry Shortcakes? It is a question Jennifer Bowles, the publisher of *Strawberryland Gazette* is often asked. "There are many contributing factors," she says. "In my opinion little girls just grew up, parents had "bought enough," and sales slacked. Very few toys have the long life like Barbie and G.I. Joe."

"For many of us Strawberry Shortcake is still "Pink Gold" and always will be," says Jolie Dickinson, an avid Strawberry Shortcake collector and big contributor to this book. Jennifer Bowles has talked to collectors aged 5 years to 75 years who all enjoy Strawberry's simple message of "Life is delicious!"

Jolie, Jennifer and I enjoyed working together on this book. We have intended for you to use it as a guide while tracking down these sought after treasures. Not all dolls, miniatures, and accessories are mentioned, but the many that are shown will give the collector some idea of what to look for! We hope you enjoy this guide "berry" much and remember that "friends are the berries!" Prices shown may vary according to area as well as demand, availability and condition.

Dolls

Strawberry Shortcake order form as a promotional for Little Debbie Shortcake Rolls. 1995. T.C.F.C. copyright. Doll manufactured by Dan Dee International Ltd.

Strawberry Shortcake rag doll. 1991. $10-20

Mail away premium from Little Debbie Cakes. Strawberry Shortcake rag doll in original box. 1996. $15-20

Raspberry Tart
stuffed, washable and
dryable rag doll with
removable outfit and
yarn hair. 15". Kenner.
1982. $20- 30

Cherry Cuddler rag doll. Kenner. $25-35
loose; $55-65 in box

Lemon Meringue stuffed, washable and dryable rag doll with removable outfit and yarn hair. 15". Kenner. 1981-82. $20-30

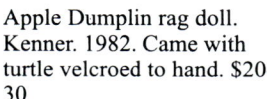

Apple Dumplin rag doll. Kenner. 1982. Came with turtle velcroed to hand. $20-30

Lime Chiffon stuffed, washable and dryable rag doll with removable outfit and yarn hair. 15". Kenner. 1983. $20-30

Orange Blossom rag doll. Kenner. 1982. $20-35

Strawberry Shortcake rag doll. Characters from Cleveland. 1991. $15-25

Apricot stuffed, washable and dryable rag doll with removable outfit and yarn hair. 13". Came with small velcroed Hopsalot. Kenner. 1982 $20-30 doll

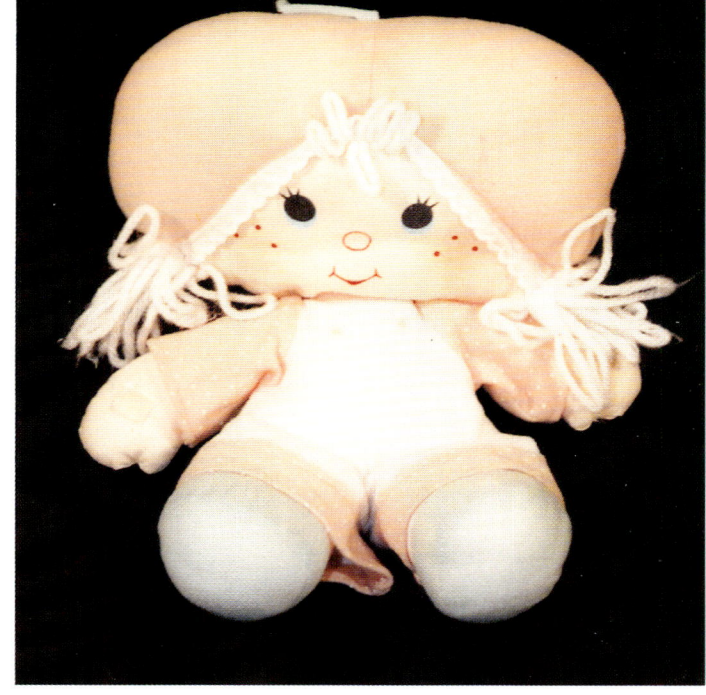

Baby Strawberry Shortcake has soft vinyl body with removable outfit. Squeeze her tummy and she blows make-believe strawberry scented kisses. Kenner. 1982. $25-40

Baby Lemon Meringue has soft vinyl body with removable outfit. Squeeze her tummy and she blows make-believe lemon scented kisses. Kenner. 1982. $25-40

Baby Apricot has soft vinyl body with removable outfit. Squeeze her tummy and she blows make-believe apricot scented kisses. Kenner. 1982. $25-40

Orange Blossom vinyl Blow-A-Kiss doll. Kenner. 1983. $25-30

"Baby Needs a Name Blow-A-Kiss" vinyl doll. Kenner. 1984. $10-15. If complete with hat, romper and booties. $25-35

Poseable Apricot with Hopsalot bunny. Kenner. 1981. $10-15

Poseable Butter cookie with Jelly Bear. Kenner. 1983. $10-15

Lemon Meringue baby doll. Kenner. 1984. $8-10

Lemon Meringue baby doll. 1984. American Greetings. $5-8

Blueberry Muffin baby doll. Kenner. 1984.
$3-6

Miniature rubber Strawberry Shortcake doll,
in a cotton stuffed baby carrier. 1984. $15-
20

Strawberry Shortcake baby doll. Kenner.
Missing boots and bottle. 1982. $5-8

Poseable Apple Dumplin with Tea Time Turtle. Kenner. 1979-80. 1st turtle made. $10-15

Poseable Cherry Cuddler with Gooseberry Goose. Kenner. 1983. $10-15

Lemon Meringue rubber jointed doll with Butter Cookie. Removable washable outfit. Came with a comb to fix their hair. Baby sold separately. 1982. Kenner. $15-20 doll; Butter Cookie $3-5

Lime Chiffon rubber jointed doll with Parfait Parrot. Removable washable outfit. Came with a comb to fix their hair. Kenner. $10-15 loose; $15-20

Cafe Ole with Burro Burrito.
International Series. Kenner. 1983.
Came as set. $15-20

Almond Tea doll with Marza
Panda. International Series.
1983. Came as set. $15-20

Crepe Suzette rubber jointed doll with removable washable outfit. Eclair the dog. Came with a comb to fix their hair. International Series. 1983. Kenner. $15-20

Blueberry Muffin doll. Sweet Sleeper. 1982-84. Kenner. $8-10 as is; $25-55 with clothes, boxed

Strawberry Shortcake flat hand doll. Kenner. 1979- 80. $10-15

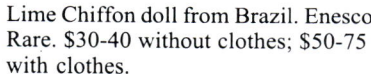

Lime Chiffon doll from Brazil. Enesco. Rare. $30-40 without clothes; $50-75 with clothes.

Strawberry Shortcake rubber head doll. Kenner. 1982- 84. Berrykin doll set with Custard pet. This doll is re-dressed. $20-25

Huckleberry Pie doll. Kenner. 1979-80. Flat hand doll, the first ones ever made of Strawberry kids. More valuable than curved hand dolls. $10-15

Mint Tulip with Marsh Mallard. Kenner. 1983. Mallard pet is from Party Pleaser doll, not from the regular doll set. $20-30

Mint Tulip with Marsh Mallard. 1983. American Greetings. $8-10

Orange Blossom, Raspberry Tart, Huckleberry Pie, Blueberry Muffin, Strawberry Shortcake are all rubber jointed dolls with removable washable outfits. Each came with a comb to fix her hair. 1981. Kenner. $10-15 each

Angel Cake with Souffle pets. Kenner. 1982. Left: Party Pleaser pet; right: regular pet. No Party Pleaser Angel Cake is shown. $15-20

Strawberry Shortcake rubber head.
doll. Characters from Cleveland.
1991. $2-4

Orange Blossom rubber head
doll. Characters from
Cleveland. 1991. $6-10

Strawberry Shortcake rubber head doll. Characters from Cleveland. 1991. $3-6

Lime Chiffon rubber head doll. Characters from Cleveland. 1991. $3-6

Lime Chiffon rubber head doll. Characters from Cleveland. 1991. $3-6

Plum Pudding Party Pleaser with Elderberry Owl. $125-150 loose; $150 and up in box

Peach Blush Party Pleaser with Melonie Belle lamb. $100 and up; Mint in box, $125 and up

Strawberry Shortcake doll and Berrykin. $125 and up

Lem and Aida with Sugarwoffer. International Series of dolls. Kenner. 1983. $25-35

Strawberry Shortcake in box with extra set of clothes. Characters from Cleveland. 1991. $15-20

Raspberry Tart in box with extra set of clothes. Characters from Cleveland. 1991. $15-20

Berry Beauty Shop. Includes paper doll and poseable doll with accessories. Those Characters from Cleveland. 1991. $15-25

Back of package for THQ dolls. 1991.

Strawberry Shortcake vinyl head rag dcll. $50-75

Vinyl head Huckleberry Pie rag doll. $50-75

Strawberry Shortcake cotton stuffed mini-pillow doll. 1982. $3-6

Cotton stuffed Strawberry Shortcake pillow doll. 19". 1982. $6-8

Plum Pudding large cotton stuffed pillow doll. 1982. $15-20

Blueberry Muffin cotton stuffed pillow doll. 1982. $10-15

Raspberry Tart cotton stuffed pillow doll.
1982. $10- 15

Apple Dumplin cotton stuffed pillow doll.
1982. $6-8

Apple Dumplin cotton stuffed pillow doll.
1982. $5-7

Apple Dumplin cotton stuffed pillow doll.
1982. $10- 15

Huckleberry Pie cotton stuffed mini-pillow
doll. 1982. $3-5

Huckleberry Pie farmer. Stuffed pillow doll.
13". 1982. $6-10

Marmalade butterfly cotton stuffed pillow doll. 1982. $4-6

Cotton stuffed Custard the cat, pillow doll. 1982. $5-6

Custard, Strawberry Shortcake's pet cat, cotton stuffed pillow doll. 1982. $5-7

Pupcake cotton stuffed pillow doll. 1982. $3-5

Sour Grapes rubber figure. Villain of Strawberryland. 1982. $20-25

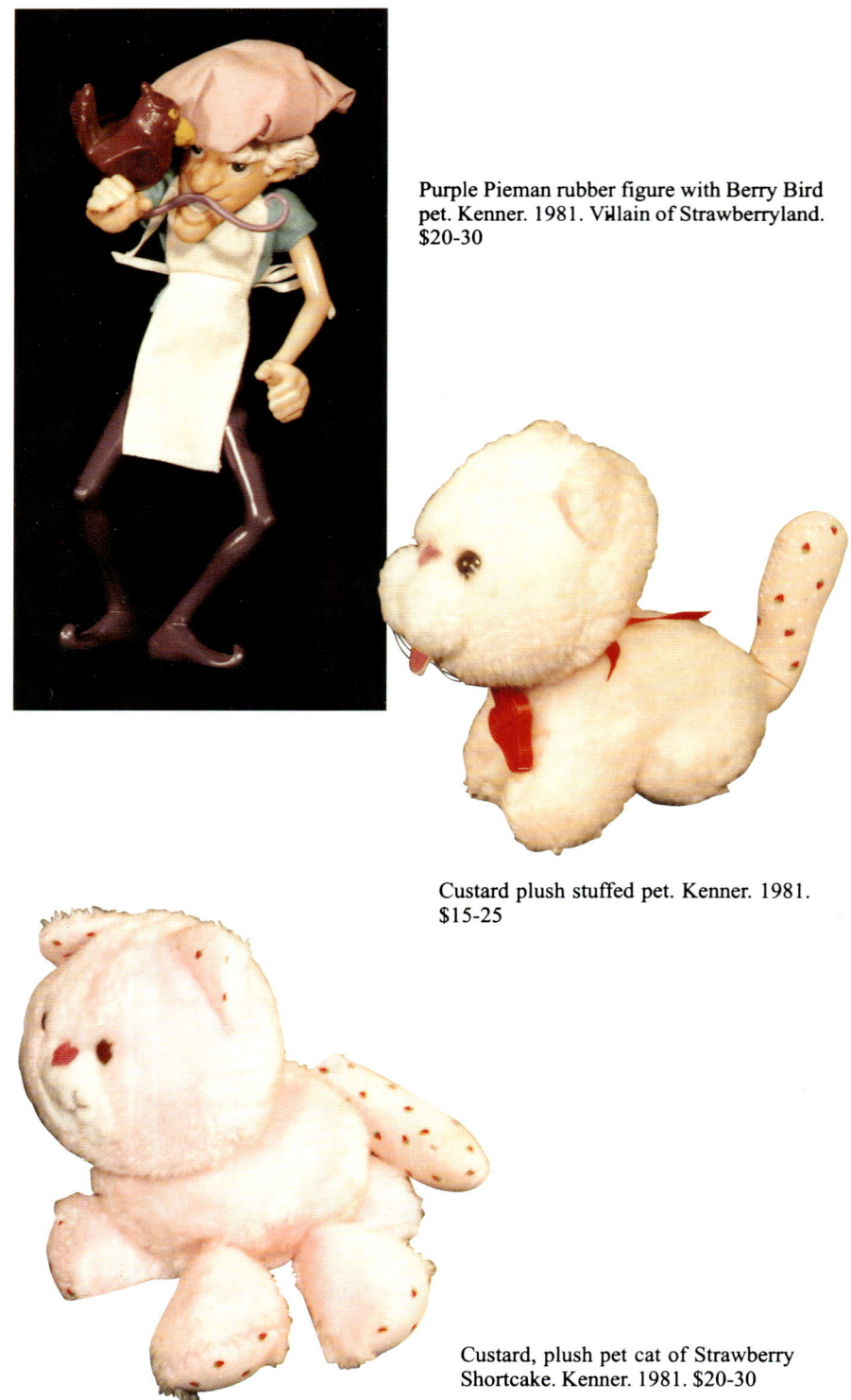

Purple Pieman rubber figure with Berry Bird pet. Kenner. 1981. Villain of Strawberryland. $20-30

Custard plush stuffed pet. Kenner. 1981. $15-25

Custard, plush pet cat of Strawberry Shortcake. Kenner. 1981. $20-30

Pupcake miniature
plush toy. Kenner.
1982. $10-20

Apricot's plush pet
Hopsalot. Kenner. 1982.
$15-20

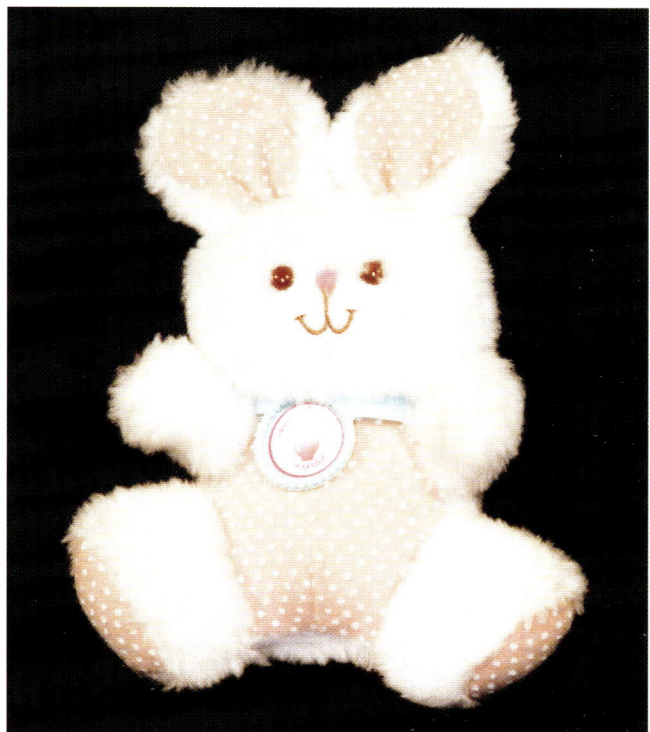

Playthings and Houses

Plastic Snail Cart in box. Kenner. 1982. $20-35

Strawberry Shortcake's plastic bicycle. Kenner. 1982. $10-15 loose; $15-20 in box.

Strawberry Shortcake plastic carousel.
Kenner. 1981. $25-35

Berry Busy Bug. Plastic toy for miniatures to ride in. Kenner. 1982. $15-25 loose; $20-30 boxed

Flitterbit the Butterfly. Plastic toy. Kenner. 1981. Missing two seats that sat on his back. $15-20

Philbert plastic caterpillar. Kenner. 1982-83. $10-20

Plastic Oats mobile. Characters from Cleveland. $30- 45

Plastic Big Berry Trolley. Kenner. 1981. $30-45

Vinyl tub toy with Strawberry Shortcake and Custard. 1982. $10-15

Playdough mold from Strawberry Shortcake Playdough Set. 1981-82. $2-3

Plastic wind up "Learn To Tell Time" play clock. 1982. $10-20. In box $25-35

Strawberry Shortcake Lite-Brite refill by Hasbro. 1982. $15-20

Strawberry Shortcake plastic windup toy. 1984. American Greetings. $10-20

Strawberry Shortcake Game Basket. American Greetings. Parker Bros. 1981. $10-15

Strawberry Shortcake Berry-Go-Round game. Parker Bros. 1981. American Greetings. $15-20

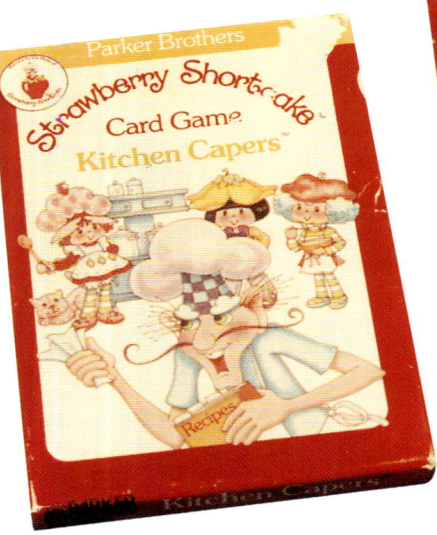

Strawberry Shortcake Kitchen Capers Card Game. Parker Brothers. 1983. $6-10

Strawberry Shortcake Playset. Rose Art.
1992. $10-15

Strawberryland game. Roseart. 1991. $10-20

Strawberry Shortcake In Big Apple City game. Parker Brothers. $5-8

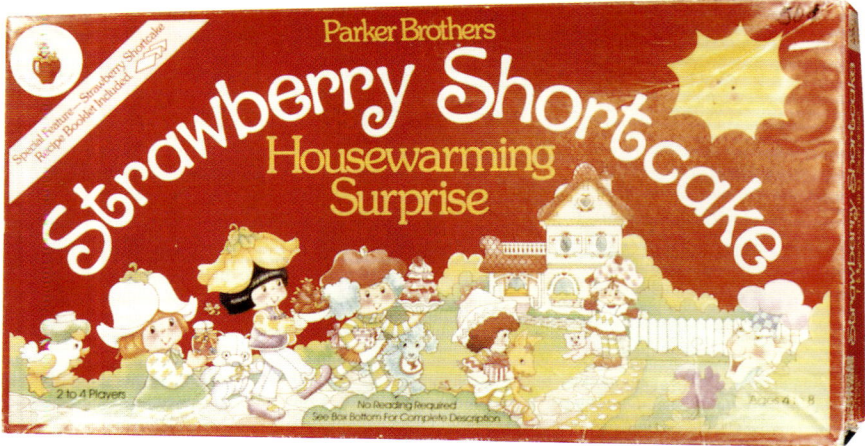

Strawberry Shortcake Housewarming Surprise game. American Greetings. Parker Brothers. 1983. $10-15

Left: Berry Snuggly Bedroom set in box. $80-100. Right: Berry Cheery Living Room set in box. 1983. $40-50

Left: Berry Cozy Kitchen in box. 1983. $40-50. Right: Berry Snuggly Bedroom in box. 1983. $35-45

Strawberry Shortcake plastic furniture. 1983. Kenner. $12-15

Berry Dainty Dining Room set in box. Kenner. 1983. $35-45.

Berry Dainty Dining Room set in box. 1983. $40-50.

Strawberry Shortcake plastic furniture. 1983. Kenner. $13-15

Strawberry Shortcake plastic furniture. 1983. Kenner. $10-15

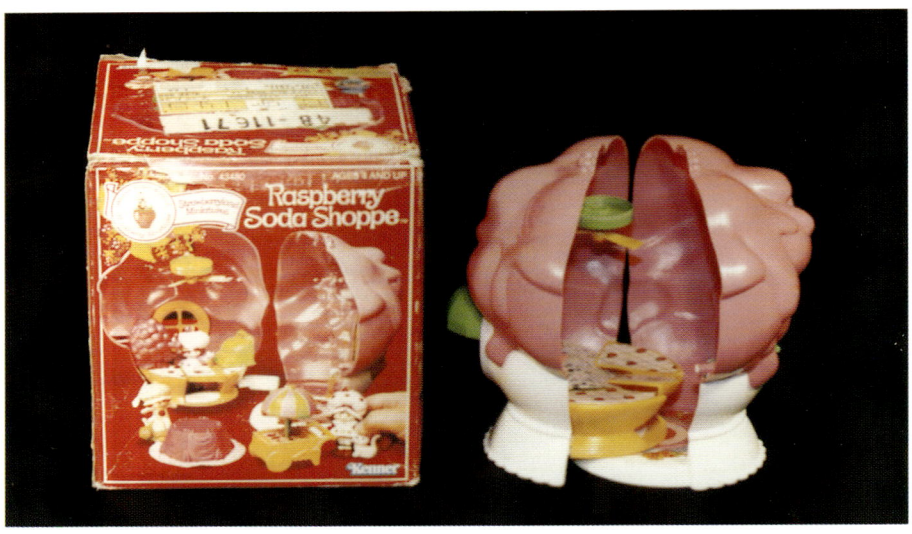

Raspberry Soda Shoppe. Kenner. 1981. $25-35

Plastic Strawberry Shortcake "Berry Patch" carrying case. All the miniatures can be stored in this case. It can be hung on the wall, sit on a table or just be carried to a friend's house. Kenner. Miniatures sold separately. 1981. $10-20 common

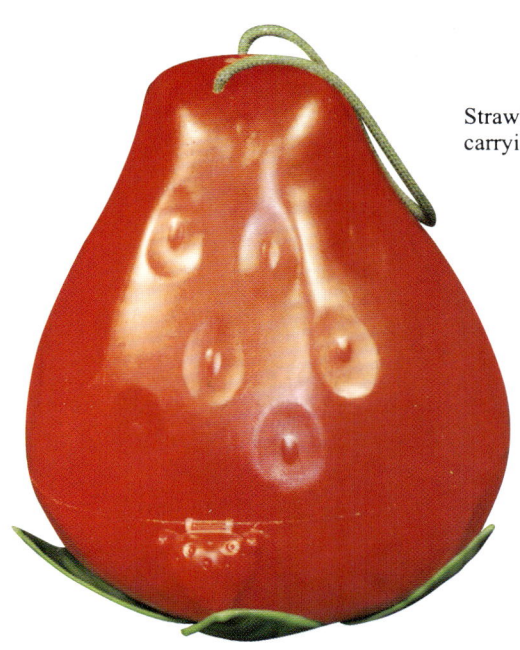

Strawberry Shortcakes plastic strawberry carrying case. Kenner. 1982. $10-15

Plastic Berry Bakeshop. Kenner. 1982. $40-50

Plastic Strawberry Shortcake mini-house. The doors and windows open and close. Figures can be packed inside when playtime is over. Figures sold separately. 1981. $35-50

Strawberry Shortcake plastic Garden Gazebo House. This vine covered gazebo came with a cupboard with hangers for keeping hats and outfits, a pretend barbecue, two chairs, a table and two hammocks. 1981. $30-45 loose; mint-in-box $50 & up

Strawberry Shortcake plastic house. 1983. Kenner. $125-200. $250-350 with furniture.

Kitchen

Plastic salt and pepper shakers. "Not toooo much" for pepper and "Just a pinch" for salt. $10-15 pair

Set of four Strawberry Shortcake juice glasses. $8-12 set (common)

Plastic cup from Deka. $6-8

Strawberry Shortcake 6 ounce glass. $3-5

Strawberry Shortcake juice pitcher. $15-20

Set of three glass Strawberry Shortcake jars. $15-20 set

Glass juice container. $8-10

Ceramic Strawberry Shortcake mug. $10-20

Plastic mug. American Greetings. Deka. 1980. $5-8

Apple Dumplin Anchor Hocking cup. $4-5

Anchor Hocking mugs. Strawberry Shortcake, Blueberry Muffin, Raspberry Tart. $8-10 each

Plastic cereal bowl. Deka. 1980. $5-10

Anchor Hocking bowl. American Greetings. $5-8

Plastic Blueberry Muffin bowl from Silite. $6-1

Plastic Strawberry plate from Deka. $4-6

Plastic Fun and Friends plate from Silite. $6-8

Plastic divided child's dish. American Greetings. 1980. Deka. $8-10

Wilton plastic cake tops. $5-10

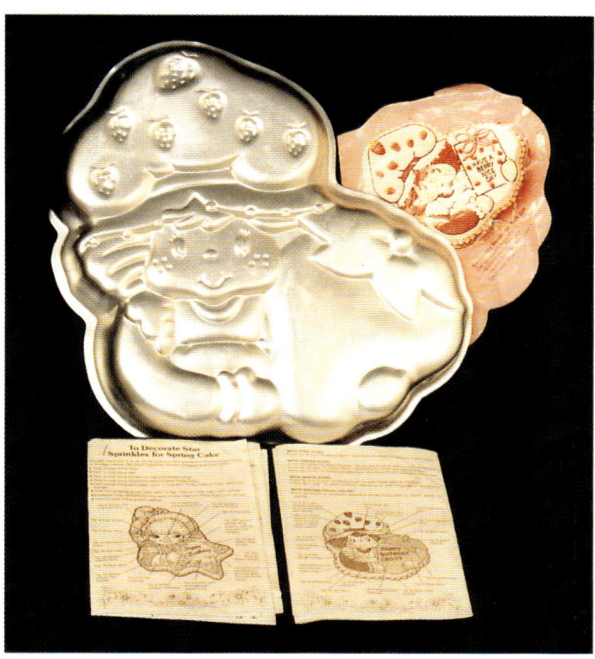

Aluminum cake pan with directions. Wilton. $15-20.

Cotton Strawberry Shortcake pot holder. $8-10

Strawberry Shortcake lunch box. $10-15

Strawberry Shortcake thermoses by Aladdin. $5-8

Metal lunch box and thermos. Berrykins. $30-40

Metal Strawberry Shortcake lunch pail. 1981. $25-35

Metal lunch box and thermos. $25-35

Plastic lunch box by Aladdin. $20-30

Strawberry Shortcake candy tin. $10-15

Mini tins. $10-15 each

Strawberry Shortcake tin container. American Greetings. $15-20

Strawberry Shortcake tin. $4-6

Tin Strawberry Shortcake container. $5-8

Metal Strawberry Shortcake tray. American Greetings. $15-25

Strawberry Shortcake metal tray. American Greetings. 1982. $20-30

Strawberry Shortcake metal tray. American Greetings. 1983. $20-30

Strawberry Shortcake metal tray. American Greetings. 1982. $20-30

Strawberry Shortcake tin cookie container.
$15-20

Hard plastic pencil holder. 1983. $15-20

Strawberry Shortcake's Cooking Fun.
Michael J. Smollin. 1980. $6-7

Tin candle container. "Life is delicious."
American Greetings. $6-10

Musical

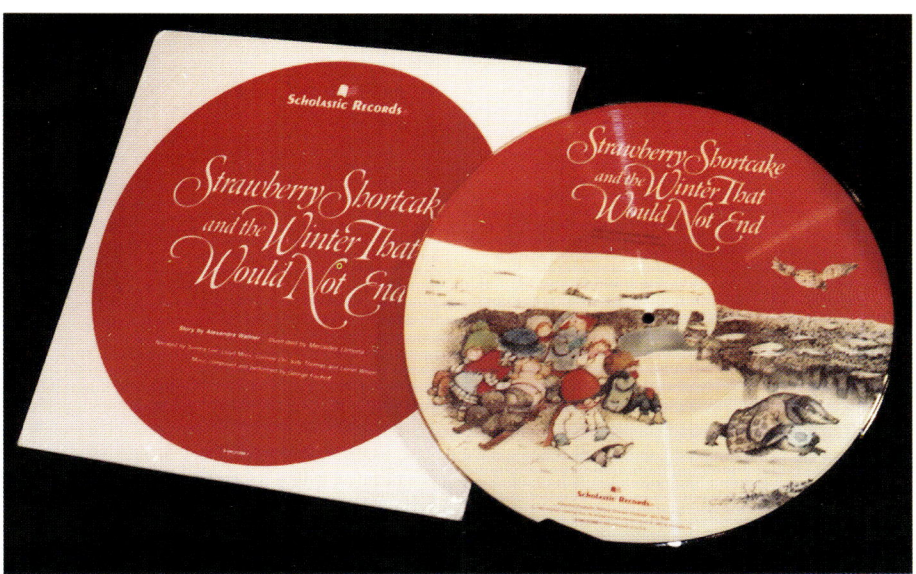

Plastic front and back of "Strawberry Shortcake and the Winter That Would Not End" record with cover. Scholastic records. 1980 promo. $15-20

Strawberry Shortcake limited edition picture disc. by Kids Stuff. 1981. American Greetings. $15-20

"Strawberry Shortcake's Alphabet Record." Kid Stuff. 1980. $8-12

"The World of Strawberry Shortcake" record. Kid Stuff. 1980. $8-12

"Strawberry Shortcake Live" record. American Greetings. $7-10

"Strawberry Shortcake Sing-a-Long" record. American Greetings. 1981. $10-15

"Strawberry Shortcake's Country Jamboree" record. Kid Stuff. American Greetings. 1980. $10-15

Strawberry Shortcake record. Limited edition. Picture disc that is numbered. Kid Stuff. American Greetings. 1981. $30-40

Strawberry Shortcake record by Kids Stuff. 1981. American Greetings. $7-10

"Strawberry Shortcake Meets the Spelling Bee." Book and record. Kid Stuff. 1981. $6-7

"Strawberry Shortcake's Day in the Country." Book and record. Kid Stuff. 1981. $4-6

"Strawberry Shortcake presents 'Apricot' & 'Hopsalot'" book and record. Kid Stuff. 1982. $5-6

"Strawberry Shortcake's Adventures in Strawberry Land." Book and record. Kid Stuff. 1980. $6-7

"Strawberry Shortcake's Book of Words," book and record. Kid Stuff. 1982. $6-7

"Strawberry Shortcake's Sweet Songs." Kids Stuff. $6-8

"Lemon Meringue & Frappé" record and book. 1982. $4-6

See, hear, and read "Strawberry Shortcake & Her Friends" book and record. 1981. $5-7

"Blueberry Muffin & Cheesecake." Kid Stuff. $3-6

"Strawberry Shortcake and the Baby Without a Name," video. 1984. $15-20

"Strawberry Shortcake Pets on Parade" video. American Greetings. $15-20

"Strawberry Shortcake Meets the Berrykins" video cassette. American Greetings. 1994. $15-20

"Strawberry Shortcake Housewarming Surprise" video cassette. Family Home Entertainment. 1994. $10-15

"The Wonderful World of Strawberry Shortcake" video. Best Video Corp. American Greetings. $12-15

Books

Christmas in Strawberryland and *Halloween in Strawberryland* story books. 1983. $3-5

Strawberry Shortcake's Outdoor Fun! story book. Michael A. Vaccaro. 1982. $10-12

Strawberry Shortcake and the Catnapping pop-up story book. 1982. $5-7

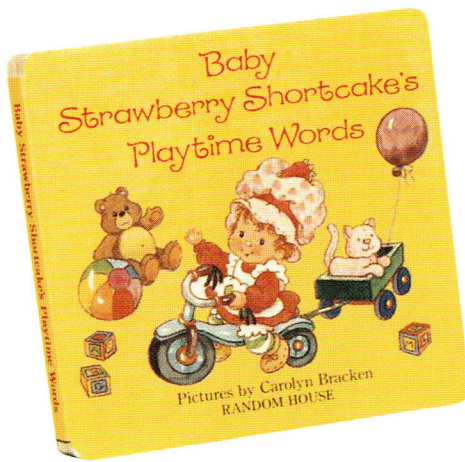

Baby Strawberry Shortcake's Playtime Words, storybook. Random house. 1984. $6-10

Strawberry Shortcake's 1-2-3 book. Randcm House. 1981. $6-10

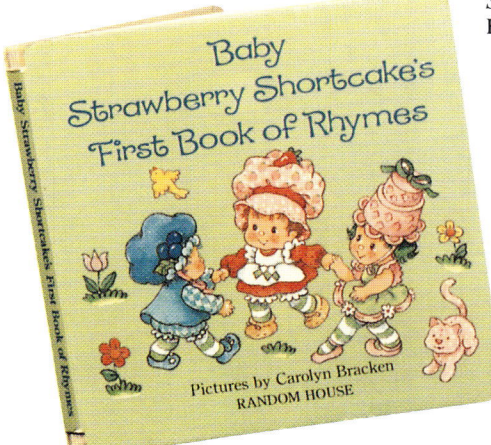

Baby Strawberry Shortcake's First Book of Rhymes. Random House. 1984. $6-10

Strawberry Shortcake and the Picnic Plot, pop-up story book. Little Pops. 1982. $6-8

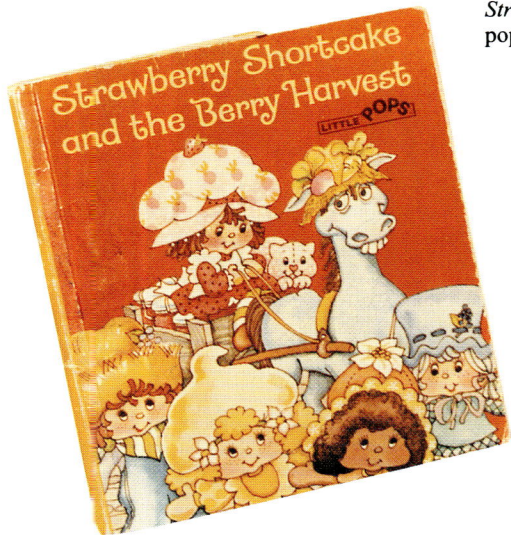

Strawberry Shortcake and the Berry Harvest pop-up story book. Little Pops. 1982. $5-7

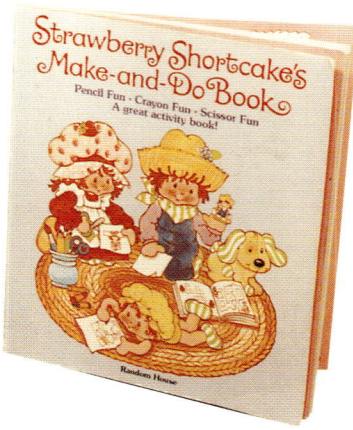

Strawberry Shortcake's Make-and-Do Book. Random House. 1980. $6-8

Strawberry Shortcake's Party Fun. Michael J. Smollin. 1983. $4-5

Strawberry Shortcake Coloring Book. Kenner. 1981. $6-8

Strawberry Shortcake and the Birthday Surprise story book. Parker Brothers. 1983. $6-8

Strawberry Shortcake and Pets on Parade story book. Parker Brothers. 1983. $5-7

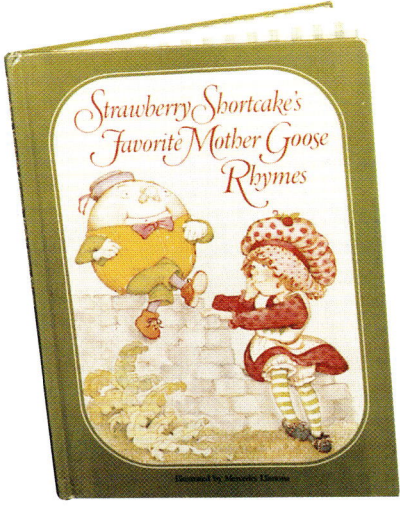

Strawberry Shortcake's Favorite Mother Goose Rhymes story book. Parker Brothers. 1983. Came in hard cover and soft. $8-10

Strawberry Shortcake and the Baby Needs a Name story book. Parker Brothers. 1984. $8-10

Strawberry Shortcake and the Big Balloon Race story book. Parker Brothers. 1983. $6-8

Strawberry Shortcake and the Deep Dark Woods story book. Parker Brothers. 1983. $6-8

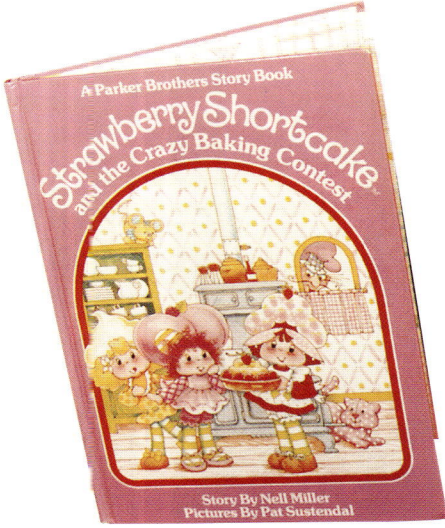

Strawberry Shortcake and the Crazy Baking Contest story book. Parker Brothers. 1983. $8-10

Baby Strawberry Shortcake: A Surprise for Baby Blueberry Muffin story book. Parker Brothers. 1984. $6-8

Baby Strawberry Shortcake: Fig Boot's Happy Day. Parker Brothers. 1984. $4-6

The Adventures of Strawberry Shortcake and Her Friends. 1980. $8-12

Strawberry Shortcake and the Winter that Would Not End story book. 1983. American Greetings. Came in soft and hard cover. $8-10

The Happy World of Strawberry Shortcake
story book. Michael A. Vaccaro. 1981. $5-10

Learning Colors with Strawberry Shortcake
story book. 1980. $6-7

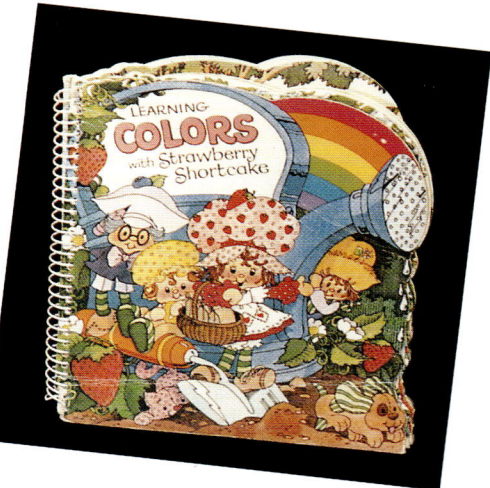

The Sweet smell of Strawberryland. Scratch
and sniff book with 8 fragrances. Random
House. 1980. $6-10

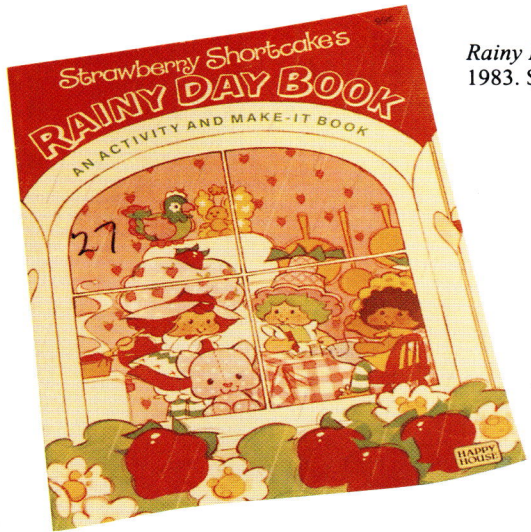

Rainy Day Activity Book. Happy House. 1983. $4-5

Color and Sew Storybook. Craft Master. 1983. $6-10

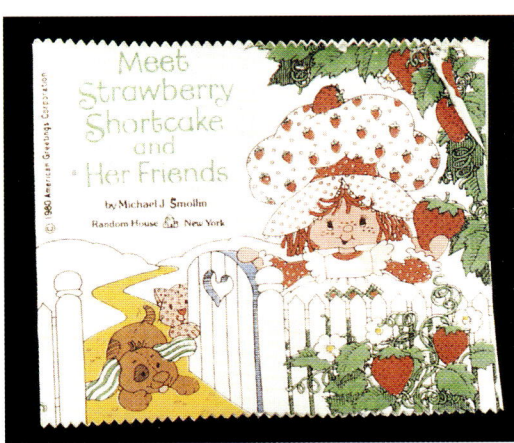

Washable cloth *Meet Strawberry Shortcake and Her Friends* story book. Random House. 1980. $6-8

PVC and Figurines

Strawberryland PVC pets that were included with dolls. 1982. $3-6 each

Mini Huckleberry Pie with Pupcake, Blueberry muffin with basket. PVC. Kenner. 1982. $5-7

Mini Orange Blossom with Marmalade, Strawberry Shortcake in her night gown, and Apricot with Hopsalot in a wheel barrel. PVC. Kenner. 1982. $3-6 each

Mimi Mint Tulip with Marsh Mallard, Lime Chiffon with Parfait Parrot, and Almond Tea with Marza Panda. PVC. Kenner. 1982-83. $4-7 each

Mini Butter Cookie with bear, Apple Dumplin with Tea Time Turtle, Apple Dumplin with wagon. Kenner. 1982. $5-7 each. First of series.

Mini Cherry Cuddler with Gooseberry, Apple Dumplin on a sled, Cherry Cuddler on a rocking horse, Strawberry Shortcake with Custard. Kenner. 1982. $3-6 each

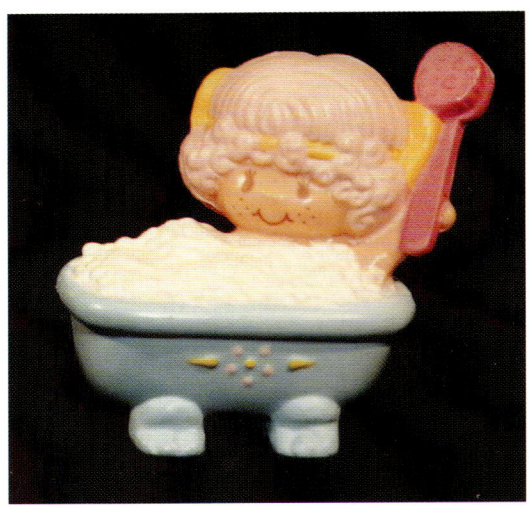

Mini Angel Cake chatting on the phone. PVC. Kenner. 1982. $2-4

Mini Angel Cake taking a bubble bath. PVC. Kenner. 1982. $3-6

Mini Apricot with Hopsalot. PVC. Kenner. 1982. $3-6

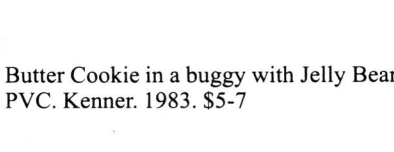

Orange Blossom with paint brush and palette. PVC. kenner. 1982. $4-6

Butter Cookie in a buggy with Jelly Bear. PVC. Kenner. 1983. $5-7

Mini Lime Chiffon ballerina and Lime Chiffon with balloons. Kenner. 1982. $3-6 each

Miniatures of Strawberry Shortcake, Raspberry Tart, Blueberry Muffin. Plastic. Kenner. 1982. $3-6 each

Miniatures of Huckleberry Pie, Mint Tulip, Strawberry Shortcake. Plastic. Kenner. 1982. $3-6 each

Miniature set of Deluxe mini figures. Kenner. 1983. $5-10 each set. Left to right-Lemon Meringue missing sewing machine, Strawberry with Custard missing a sailboat, Angel Cake missing desk. There are 12 playsets in this series.

Mini Lemon Meringue in different poses. PVC. Kenner. 1982. $3-6 each

Mini Strawberry Shortcake with bushel basket, with watering can, and with Custard. PVC. Kenner. $3-6 each

Mini Raspberry Tart with ice cream and with bowl of berries. PVC. Kenner. 1982. $3-6 each

Mini Purple Pieman. PVC. Kenner. 1982. $3-6

Mini Sour Grapes. PVC. Kenner. 1982. $8-12

Strawberry Shortcake plastic figurine. Came in painting set. 1983. $4-6

Blueberry Muffin plastic figurine. Came in painting kit. 1983. $4-6

Ceramic top of treasure box. American Greetings. $10-20

Paper, Party Goods, and Puzzles

Thank you card included in box with doll.
$1-2

Thank you post card. Included with the toy in the factory box. $1-2

Thank you post card. International series. $1-2

Strawberry Shortcake mini memo by American Greetings. Included paper sheets decorated with Lucky Bug. 1981. $3-5

Berry Baby Album came with Blow Kiss dolls. American Greetings. 1984. $3-4

Strawberry Shortcake Valentine cards. 4 different designs. American Greetings. 1981. $10-15

Strawberry Shortcake birthday card.
Characters from Cleveland. 1996. $3-4

Strawberry Shortcake birthday card.
Characters from Cleveland. 1996. $3-4

Strawberry Shortcake Valentines. 1981. $1-2

Strawberry Shortcake and friends wrapping paper. American Greetings. 1982. $8-15

Bag of Strawberry Shortcake Valentines. Plus Mark. 1983. $8-12

Strawberry Shortcake Gift Wrap. 1982. $4-5

Strawberry Shortcake table cover. American Greetings. $3-5

Strawberry Shortcake table cover. American Greetings. $3-5

Paper party table cover. American Greetings. 1985. $3-5

Package of gift tags from American Greetings. 1982. $5-7

Strawberry Shortcake decorations. Set of4 Honeycomb. 1982. American Greetings. $7-12

Strawberry Shortcake stickers. Forget-Me-Not. Re- released by Characters from Cleveland. $4-6

Variety of Strawberry Shortcake scented stickers. 1983. $6-7 each

Variety of Strawberry Shortcake scented stickers. 1983. $6-7 each

Variety of Strawberry Shortcake scented stickers. 1983. $5-6 each

Strawberry Shortcake note pad with plastic holder. 1983. $5-7

Strawberry Shortcake board. American Greetings. Freelance Inc. 1982. $4-5

Strawberry Shortcake party bag. 1980. American Greetings. $5-7

Set of 10 Holiday and Party hanging lights. 1980. $40-55

Strawberry Shortcake costume with mask.
Ben Cooper. 1983. $10-20

Cardboard 100 piece Strawberry Shortcake
puzzle. T.C.F.C. Rose Art Brand. 1992. $6-
10

Strawberry Shortcake puzzle. 1983. Craft master. $8- 10

Strawberry Shortcake puzzle. Characters from Cleveland. 1992. $5-7

Satin "Huckleberry Pie" Christmas
ornament. $20-25

Satin Strawberry Shortcake gang ornament.
1982. American Greetings. $15-20

Strawberry Shortcake candle. 3". American Greetings. 4-6

Strawberry Shortcake square candles. 1981. $4-6 each

Clothes and Accessories

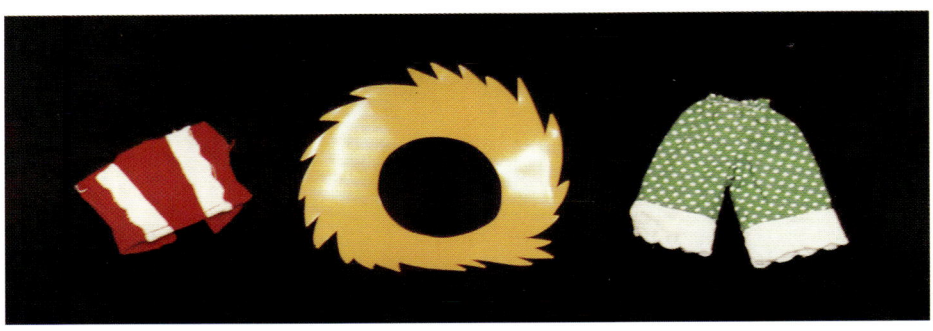

Strawberry Shortcake garden set of clothes. 1981-82. Kenner. $5-7

Strawberry Shortcake clothing. Characters from Cleveland. 1991. $2-3

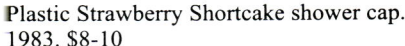
Plastic Strawberry Shortcake shower cap. 1983. $8-10

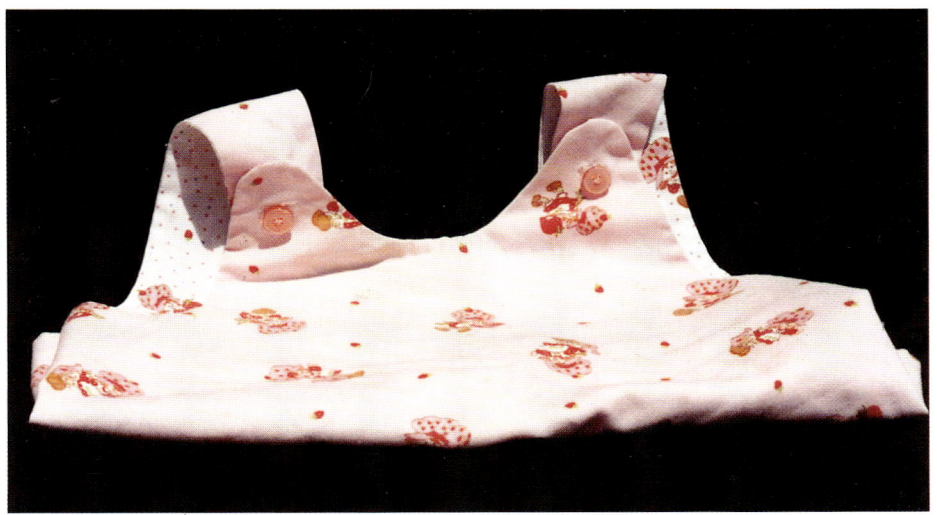

Strawberry Shortcake cotton jumper. 1982. $10-15

Nylon synthetic night gown. 1983. $6-10

Stride Rite cardboard shoe box by Stride Rite. $10- 15

Strawberry Shortcake shoes by Stride Rite.
No date. $7-10

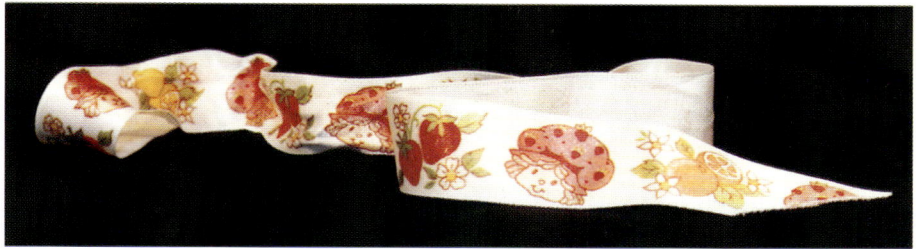

Strawberry Shortcake ribbon. American Greetings. $5- 10

Strawberry Shortcake "Life is the berries" ribbon. American Greetings. $5-8

Strawberry Shortcake enamel necklace. Send away award for "berry points" from specially marked boxes. 1982. $10-15

Strawberry Shortcake imitation pin. $3-5

Strawberry Shortcake enamel pin. Send away award for "berry points" from specially marked boxes. 1982. $10-15

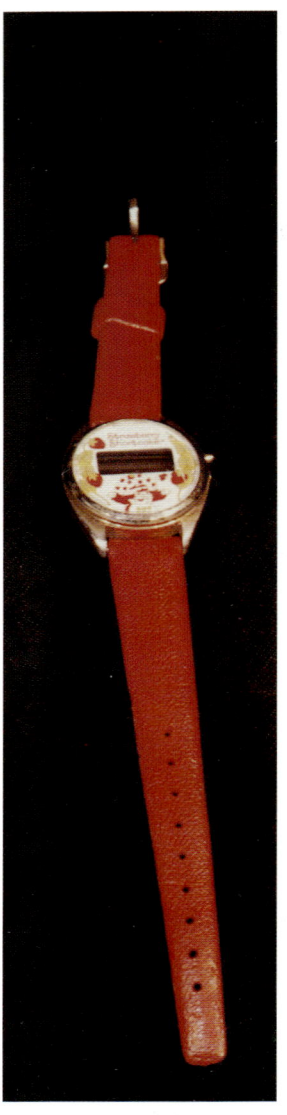

Strawberry Shortcake wristwatch. American Greetings. 1982. $15-20

116

Welcome to the World of Strawberry
Shortcake pin back button. American
Greetings. 1980. $5-8

Strawberry Shortcake trinket box. Enesco.
$6-10

Strawberry Shortcake suitcase. American Greetings. 1982. Also came in blue. $10-20

Fabric Strawberry Shortcake child's purse. $5-7

Fabric Strawberry wallet. American Greetings. $4-6

Strawberry Shortcake mirror. American Greetings. 1981. $10-15

Strawberry Shortcake hair clips.
American Greetings. 1981. $5-7

Blueberry Muffin hair clips. American
Greetings. 1981. $5-7

Blueberry Muffin plastic music box. Plays "Love Story." $40-60

Lemon Meringue divided make-up carrying basket. Includes comb, mirror, and brush. $30-40

Strawberry Shortcake divided make-up carrying basket. Includes comb, mirror and brush. $30-40

Bed and Playroom

Strawberry Shortcake picture. $6-10

Strawberry Shortcake painted glass pictures. Lu Leis. 1980. American Greetings. $12-15

Plastic frame dimensional
Strawberry Shortcake
pictures. Lu Leis. 1980.
American Greetings. $12-15
Pair

Plastic frame Strawberryland picture. 1980. American Greetings. Pro Arts Inc. $15-20

Plastic frame mirror. $5-8

Strawberry Shortcake picture frame. American Greetings. $5-7

Plastic Strawberry Shortcake switch cover. $10-15

Ceramic Strawberry
Shortcake night light.
12". 1981. American
Greetings. $35-50

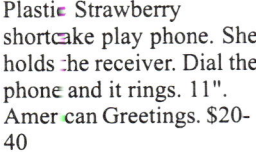

Plastic Strawberry
shortcake play phone. She
holds the receiver. Dial the
phone and it rings. 11".
American Greetings. $20-
40

Strawberry Shortcake alarm clock. American
Greetings. $30-50

Plastic Strawberry Shortcake stove and oven. Kenner. 1983. $10-20

Hardboard dressing
table with mirror.
1982. $50-75

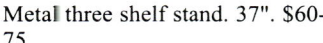Metal three shelf stand. 37". $60-
75

Hardboard Strawberry Shortcake cabinet, closed and opened. 1982. American Greetings. $50-75

Hardboard Strawberry Shortcake shelf unit. 1983. American Greetings. $50-75

Hardboard Strawberry Shortcake doll bed and high chair. Bed $40-50; chair $40-55

Common cotton sheet. $6-10

Fiberboard grow chart. Came in two pieces.
$40-50

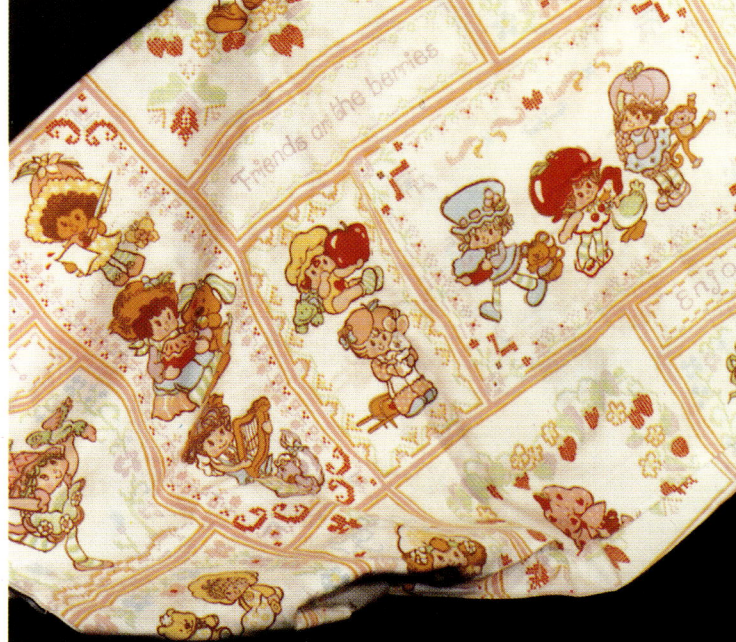

Cotton Strawberryland sheet. International Design. American Greetings. 1983. $13-15

Cotton Strawberry Shortcake sheet. American Greetings. 1982-83. $10-12

Cotton Strawberryland sheet.
American Greetings. 1982.
$15-20

Cotton flannel Straw-
berry Shortcake sheet.
Big Apple City. 1981.
American Greetings.
$15-20

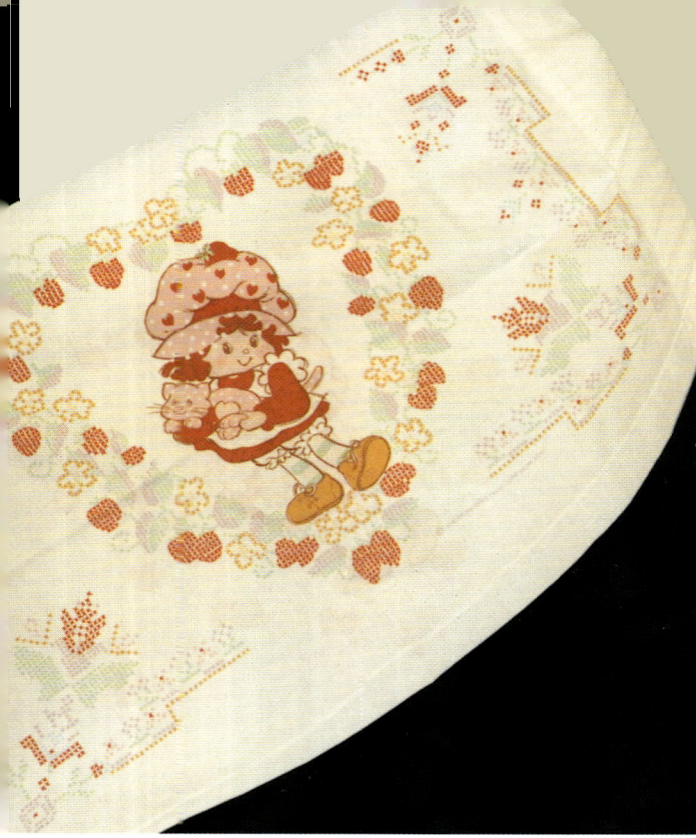

Strawberry Shortcake cotton
pillow case. 1982-83. $4-6

Cotton sheet. Mostly
common. 1981. $6-8

Cotton Strawberry Shortcake pillow case. American Greetings. 1981. $3-4

Cotton Strawberry Shortcake pillow case. American Greetings. 1981. $4-6

Strawberry Shortcake heart design sheet. American Greetings. 1982. $10-12

Cotton Strawberryland pillow case. Garden pattern. American Greetings. 1982. $4-6

Cotton Berrykins pillow case. American Greetings. 1984. $4-6

Cotton Strawberry Shortcake pillow case. 1982. Garden pattern. $4-6

Strawberry Shortcake cotton quilt.
American Greetings. 1982. $30-40

Cotton Strawberry Shortcake pillow
case. Big Apple City. 1981. $4-6

Cotton Strawberry Shortcake ruffled bedspread. American Greetings. 1982. $15-20

Cotton Strawberry Shortcake ruffled bedspread. 1980- 81. $12-15

Cotton Strawberry Shortcake ruffled pillow sham. American Greetings. 1980-81. $4-6

Cotton ruffled bedspread with Strawberry Shortcake and friends. $15-20

140

Nylon Strawberry Shortcake sleeping bag/quilt. 1982. American Greetings. $30-40

Stuffed cotton pillow. Cherry Cuddler with Gooseberry pet. No date. No copyright. $5-7

Strawberry Shortcake cotton pillow. Sewn applique. Front and back. 1982 American Greetings. $4-6

Strawberry Shortcake cotton stuffed pillow with Lucky Bug on top. $4-6

Two wheel bike with training wheels. 1982. $60-75

Bibliography

Davis, Kerra. "The Enchanting world of Strawberry Shortcake." *Toy Shop Magazine.* October 11, 1996.

Strawberryland Gazette. 405 E. Main, Greenville, Kentucky. 42345